Pond Plants

Ernestine Giesecke

Heinemann Library
Des Plaines, Illinois

Published by Heinemann Library,
an imprint of Reed Educational & Professional Publishing.
1350 East Touhy Avenue, Suite 240 West
Des Plaines, IL 60018

Designed by Depke Design
Illustrations by Eileen Mueller Neill
Printed in Hong Kong

03 02 01 00 99
10 9 8 7 6 5 4 3 2 1

Library of Congress Cataloging-in-Publication Data

Giesecke, Ernestine, 1945-
    Pond plants / Ernestine Giesecke.
        p. cm. – (Plants)
    Includes bibliographical references (p.   ) and index.
    Summary: Describes how various plants adapt to living in a pond, including the cattail, duckweed, and bladderwort.
    ISBN 1-57572-826-5 (lib. bdg.)
    1. Pond plants—Juvenile literature. [1. Pond plants.]
I. Title. II. Series: Plants (Des Plaines, Ill.)
QK938.M6G54 1999
581.763'6—dc21   98-44522
                                                        CIP
                                                        AC

Acknowledgments:

The Publisher would like to thanks the following for permission to reproduce photographs:
Cover: Dr. E.R. Degginger/Earth Scenes
Dr. E.R. Degginger/Earth Scenes pp. 4, 27; Dr. E.R. Degginger pp. 8, 10-12, 20-21, 24, 26, 28; Breck P. Kent/Earth Scenes p. 9; Bates Littlehales/Animals Animals p. 13; Robert Maier/Earth Scenes p. 14; Lee Rue III/Animals Animals p. 15; Adrienne T. Gibson/Earth Scenes p. 16; Gerald Corsi/Visuals Unlimited p. 18; Oxford Scientific Films/Earth Scenes p. 19; G.I. Bernard/Earth Scenes p. 22; V.P. Weinland/Photo Researchers p. 23; G.I. Bernard/ Animals Animals p. 25.

Some words are shown in bold, **like this.** You can find out what they mean by looking in the glossary.

CAUTION!

Whenever you study a pond, be sure you are with an adult. Be careful near the water. Leave all the plants and animals in the same place you found them.

Contents

The Pond. 4
Pond Plants 6

Pond Edge Plants
Cattail 8
Reed 10
Rush 12
Water Lily 14

Floating Plants
Duckweed 16
Water Buttercup 18
Algae 20

Underwater Plants
Pondweed 22
Bladderwort 24
Hornwort. 26

The Pond's Future. 28
Glossary *30*
Parts of a Plant. *31*
More Books to Read *31*
Index *32*

The Pond

A pond is a small body of water. Usually it is so **shallow** that light reaches the bottom.

Pond water is warm and has lots of
nutrients, the things plants need to grow.
Ponds are home to many different kinds
of plants and animals.

Pond Plants

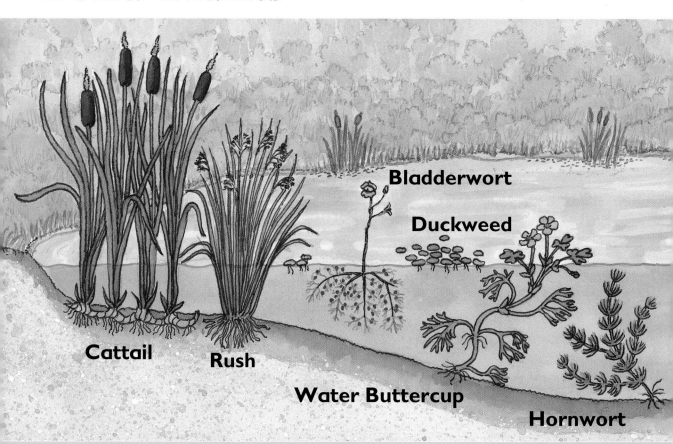

Bladderwort

Duckweed

Cattail

Rush

Water Buttercup

Hornwort

Plants are an important part of a pond. A pond must have plants before animals can live there. The plants provide food for the plant-eating animals.

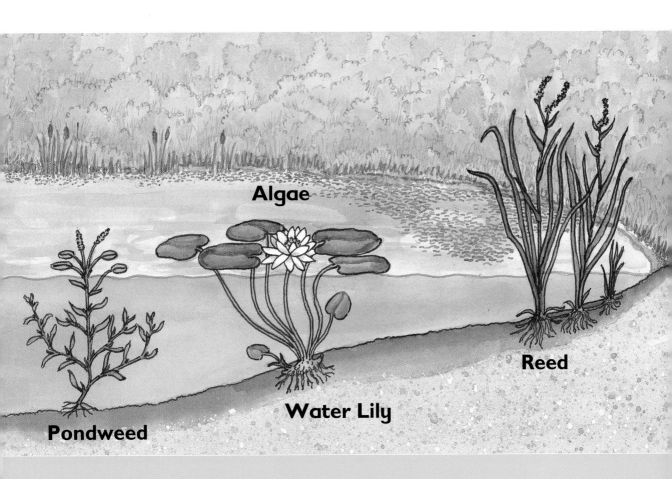

Algae

Reed

Water Lily

Pondweed

Different plants live in different parts of
the pond. Some live at the edge of the
pond. Some plants float on the **surface.**
Other plants live under the water.

Cattail

Cattails are one kind of plant you can find at the edge of a pond. Their roots are in the mud. They help trap **soil** along the shore of the pond.

The flower of the cattail plant is long and brown. It looks a little like a sausage. Some animals, like muskrat, like to eat cattails.

Reed

Reeds grow along the edges and shallow parts of a pond. Their roots and stems creep along the ground. They keep the reeds from blowing over.

A **reed** is a kind of grass. It has a round, hollow stem like a drinking straw. The reed has a feathery flower. Some birds hide their nests in reeds.

Rush

A rush is another plant that is like grass. It has a round stem and stiff, narrow leaves. Many rushes have stems filled with white fluffy **pith**. The pith helps reeds stand straight and tall.

Rushes can grow very close together.
They offer a safe place for birds to nest.

Water Lily

Water lily leaves rest on the **surface** of the pond. The water lily stem and roots stretch to the bottom.

The stem of a water lily is long and bends easily. This lets the water lily leaves—and the frog—stay on the water surface even when the water rises or falls.

Duckweed

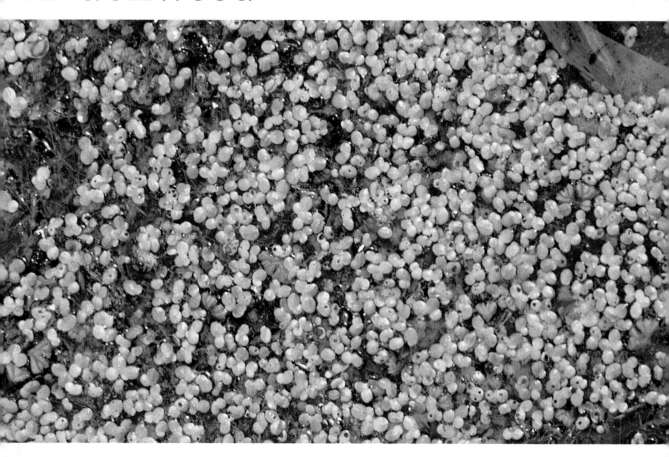

Duckweed floats on the **surface** of the pond. It is one of the smallest and simplest plants there is. Ducks and other water birds eat duckweed.

Duckweed grows quickly and can cover an entire pond. Each tiny plant **divides** itself every few days.

Water Buttercup

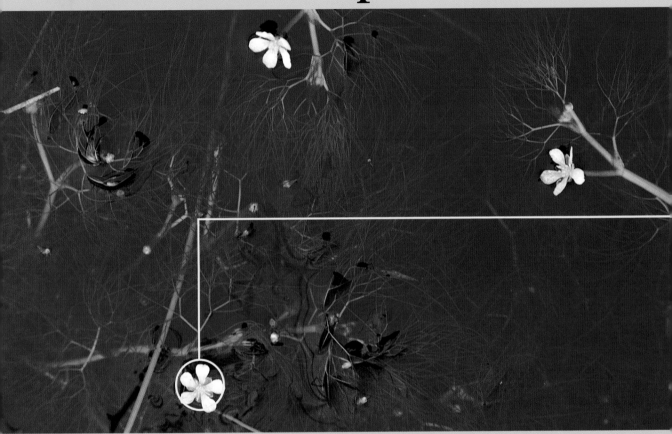

This water buttercup floats freely in the pond. The flowers can be yellow or yellow and white. The water buttercup has two kinds of leaves. The leaves above the water are rounded.

The leaves under the water are finely divided and **branched.** They help the plant get the **nutrients** it needs from the water in the pond. The water buttercup has no roots.

Algae

This gooey green mass is called **algae** (AL-jee). Algae do not have stems, leaves, or roots. Even so, this algae is like most plants—it is green and can make its own food.

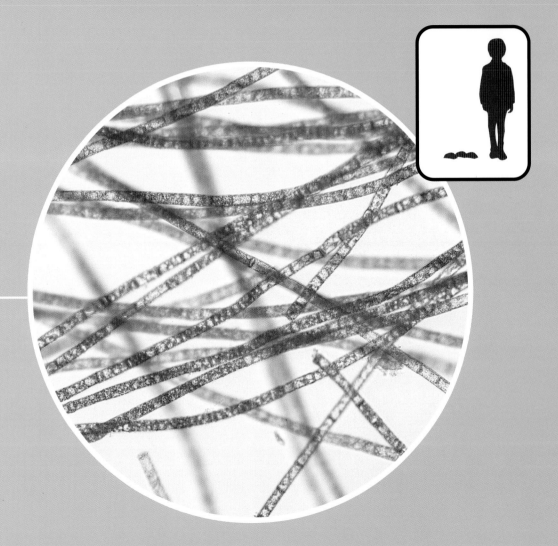

Algae looks different from plants under a
magnifying glass. It is made of millions of
tiny green threads.

Pondweed

This pondweed, like other underwater plants, puts **oxygen** into the water. Fish, snails, frogs, and other pond animals need oxygen to live.

Pondweed flowers grow in groups on the top of the plant's stem. The flowers stick out of the water while the rest of the plant stays underwater.

Bladderwort

The bladderwort has fine leaves under the surface of the water. Above the water the plant has tiny round bladders or sacs.

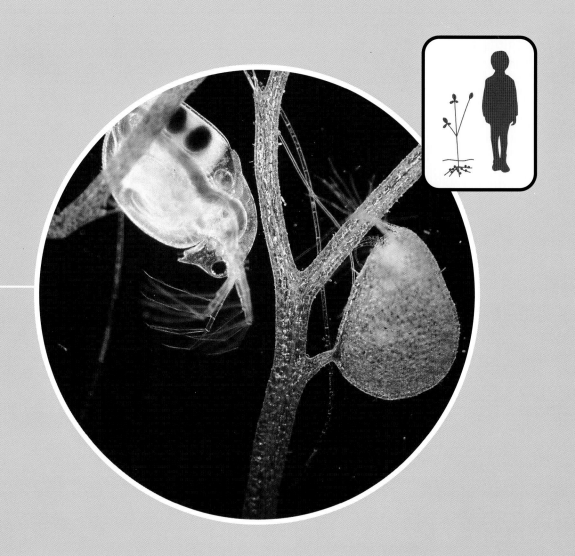

When an insect wanders into a sac, the
tiny bladder has a trap door that closes
tight. Once the door is closed, the plant
begins to **digest** the animal.

Hornwort

The hornwort is another plant that lives most of its life beneath the water. The plant produces **pollen** packets which float to the **surface** of the water.

At the surface, the packets burst and the pollen flows out. The pollen sinks until it reaches the underwater flowers of another hornwort. Then the plant makes the seeds for new plants.

The Pond's Future

Many ponds are in danger of dying.
Pollutants from factories, **pesticides**, and
fertilizers can run into ponds when it rains.
They help the algae grow very fast. It covers
the pond so sunlight can't get through.

You can help keep ponds alive and well.
Do not dump anything in a pond. Groups
in your area may work to clean the pond
scum from ponds.

Glossary

branched having side stems growing from a main stem

digest to make food ready to be used by the body

divide one thing splits into two or more pieces

fertilizers things people put on plants to make them grow faster

nutrients things plants and all life needs to grow

oxygen a gas in air needed by animals to live

pesticides things that people put on plants to kill insects that might eat the plants

pith soft spongy material inside some plant stems

pollen dust-size grains that a plant needs to make seeds

pollutants things made by people that make the earth and living things unhealthy

shallow not deep

soil the ground a plant grows in

surface the top of something, like the top of a box

Parts of a Plant

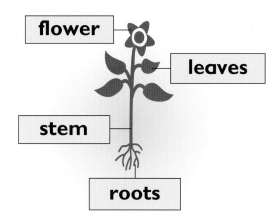

More Books to Read

Fleisher, Paul. *Pond*. Tarrytown, NY: Marshall Cavendish Corporation. 1998. An older reader can help you with this book.

Fowler, Allan. *Life In a Pond*. Danbury, CT: Franklin Watts. 1998.

Hester, Nigel. *The Living Pond*. Danbury, CT: Franklin Watts. 1990. An older reader can help you with this book.

Plants Feed on Sunlight. Brookfield, CT: Milbrook Press, Inc. 1998.

Index

algae 20–21

bladderwort 24–25

cattail 8–9

duckweed 16–17

flowers 9, 11, 23,
  27, 31

hornwort 26–27

leaves 12, 14–15, 18–19,
  24, 31

nutrients 5, 19

oxygen 22

pollen 26–27

pondweed 22–23

reed 10–11

roots 8, 10, 14, 31

rush 12–13

seeds 27

stems 10, 12, 14–15,
  23, 31

water buttercup 18–19

water lily 14–15